In the Sea

Characters

 Red Group

 Blue Group

 Purple Group

 All

Setting The sea

by Cynthia Swain

My Picture Words

clam

crab

snail

starfish

My Sight Words

is	like
little	see
the	we

We see the  .

starfish

 The  is little.

starfish

 We like the .

starfish

We see the  .

clam

 The  is little.

clam

 We like the  .

clam

We see the  .

crab

 The  is little.

crab

 We like the  .

crab

We see the 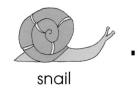 .

snail

 The 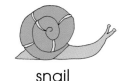 is little.

snail

 We like the 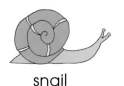 .

snail

We see the  !

whale

The End